Fuse

Basic guidelines on the component
of fuse and its uses

Dr Walt wade

Contents

Chapter1...................3

  Introduction to component of fuse..3

Chapter2..................14

  Types of fuse..........14

Chapter3..................24

  Uses of fuse............24

The end 34

Chapter 1

Introduction to component of fuse

A fuse is an electrical safety device that is used to protect electrical circuits from excessive current flows. It is a crucial component of electrical systems as it helps prevent damage to electrical equipment and can potentially save lives. In this article, we will provide an introduction to the components of a fuse, how they work, and their importance in electrical safety. Composition of a Fuse: The basic components of a fuse include a fuse element, a fuse holder, and a fuse cap. The fuse element is typically made of a thin strip of metal such as copper, silver or aluminum. This element is designed to melt when high levels of current flow hthrough it, breaking the circuit and protecting the rest of the system. The fuse holder is the part that holds the fuse element in place and is usually made of a non-conductive material such as porcelain, glass, or plastic. It is important that the material is non-conductive to prevent the flow of electricity from the fuse element to the surroundings. The fuse cap is the metal end of the fuse that connects the fuse element to the circuit. It is usually made of brass, copper or aluminum to ensure good conductivity. Some modern fuses also include an indicator light on the fuse cap to show if the fuse has blown. Working Principle of a Fuse: The main purpose of a fuse is to limit the flow of current in a circuit by melting the fuse element when there is an excessive or abnormal flow of electricity. This is achieved through the use of a fuse with a predetermined current rating specific to the electrical system it is protecting. When the current exceeds this rating, the fuse element will melt, breaking the circuit and preventing any further damage. The melting of the fuse element is caused by the high level of joule heating, also known as resistance heating. As current flows through the small strip of metal, it encounters resistance which generates heat. This heat increases with the increase in current until it reaches a point where it melts the fuse element and breaks the circuit. Types of Fuses: Fuses come in various types, each designed for different applications and environments. Some of the most common types of fuses include: 1. Cartridge Fuses - These are cylindrical fuses with metal contacts on either end. They are commonly used in electrical systems with high current ratings. 2. Blade Fuses - As the name suggests, these fuses have a flat, blade-like design and are found in most modern cars. They are mostly used in low voltage applications. 3. Glass Tube Fuses - These fuses have a similar design to a light bulb and are commonly used in household appliances. 4. Thermal Fuses - These fuses are designed to blow at a certain temperature, protecting electrical equipment from overheating. Importance of Fuses in Electrical

Systems: Fuses are an essential component of electrical systems as they play a vital role in protecting against electrical fires and other hazards. They help to regulate the flow of electricity, prevent damage to equipment and ensure the safety of individuals. Some of the key benefits of fuses include: 1. Overload Protection - Fuses are designed to blow and interrupt the flow of electricity when there is too much current. This helps prevent damage to electrical equipment and ensures the safety of the users. 2. Cost-Effective - Fuses are relatively inexpensive when compared to other forms of circuit protection such as circuit breakers. This makes them a cost-effective option for protecting electrical systems. 3. Quick

Response Time - Fuses have a fast response time, typically blowing within milliseconds when an electrical fault occurs. This quick response helps minimize the potential damage to equipment and protects against fire hazards. 4. Easy to Replace - When a fuse blows, it can be easily replaced by simply removing the old one and inserting a new one. This makes them convenient and user-friendly for maintenance and repairs.

Fuse wire

Fuse wire, also commonly known as safety wire or electrical fuse, is a thin, metallic wire that is designed to melt and break when exposed to excessive electrical current. This mechanism of fuse wire is crucial in preventing electrical fires and damage to electrical systems. Fuse wire is used in various electrical applications, such as household wiring, electrical appliances, and industrial equipment. In this article, we will delve deeper into the world of fuse wire, its types, uses, and benefits. Types of Fuse Wire: There are mainly three types of fuse wire: quick blow, time-delay, and ultra-rapid fuse wires. 1. Quick Blow Fuse Wire: Quick blow fuse wire is designed to melt quickly when exposed to high levels of electrical current. These types of fuse wires are used in applications where immediate protection is required, such as in wall sockets or light fixtures. Quick blow fuse wires have a small time delay before they melt, but not significant enough to cause any damage. 2. Time-Delay Fuse Wire: Time-delay fuse wire, as the name suggests, has a controlled delay time before it breaks. This type of fuse wire is commonly used in electronic devices such as televisions, computers, and stereos. The delay in melting allows the device to function normally during power surges or momentary increases in current. Time-delay fuse wire also has a higher current rating compared to quick blow fuse wire. 3. Ultra-Rapid Fuse Wire: Ultra-rapid fuse wires are designed to melt almost instantly when exposed to excessive levels of current. They are commonly used in industrial settings where the equipment requires high levels of current to function. These fuse wires provide maximum protection against electrical fires and damage to equipment. Uses of Fuse Wire: Fuse wire is an essential component in any electrical system. It is used in a variety of applications, including household and industrial wiring, electrical appliances, and electronic devices. Following are some common uses of fuse wire: 1. Household Wiring: Fuse wire is used in household wiring to protect the electrical system from power surges. It is installed in a fuse box, also known as a fuse panel, which is a central point that distributes electricity to different parts of the house. If the system is overloaded, the fuse wire will melt, breaking the circuit and preventing potential hazards. 2. Electrical Appliances: Most electrical appliances, such as refrigerators, microwaves, and televisions, have fuse wires installed in them. These fuse wires offer an extra layer of protection in case of power surges, short circuits, or other electrical faults. 3. Industrial Equipment: In industrial settings, fuse wire is used to protect machinery and equipment that require high levels of current to function. The melting of fuse wire alerts the operators to any potential electrical issues, allowing them to take necessary precautions to prevent accidents. Benefits of Fuse Wire: 1. Protects Against Electrical Hazards: Fuse wire is designed to protect against electrical hazards such as fires, explosions, and damage to equipment. In case of a power surge or overload, the fuse wire melts, breaking the circuit and preventing the electrical system from damaging. 2. Cost-Effective: Fuse wire is a cost-effective way to protect electrical systems. Compared to other forms of protection, such as circuit breakers or surge protectors, fuse wire is inexpensive and easy to replace. 3. Quick and Easy Replacement: If a fuse wire breaks or melts, it can be quickly and easily replaced. Unlike other forms of protection, which may require professional assistance, replacing a fuse wire is a simple DIY task.

Chapter2
Types of fuse

1. Cartridge Fuse Also known as a "barrel" fuse, this type of fuse consists of a glass or ceramic tube with metal caps on both ends. Inside the tube, there is a metal strip or wire that will melt and break the circuit when there is excessive current flow. Cartridge fuses are available in different sizes, shapes, and amp ratings, making them suitable for a wide range of applications, from household appliances to industrial machinery. 2. Blade Fuse Blade fuses, also known as "spade" fuses, have a similar design to cartridge fuses but with a plastic body and two metal prongs that plug into spring-loaded terminals. They

are commonly used in vehicles to protect automotive electrical circuits as they are compact and have a lower profile compared to cartridge fuses. 3. Thermal Fuse This type of fuse operates based on temperature change. It consists of a small resistor connected to a thermal cutoff. When the temperature exceeds its rated capacity, the resistor will expand and break the circuit, thus protecting the circuit from overheating. Thermal fuses are commonly used in hair dryers, irons, and other heat-producing devices. 4. Time-Delay Fuse Time-delay fuses, also known as "slow-blow" fuses, have a built-in time delay that allows for a temporary overload without blowing the fuse. They are commonly used in circuits with high inrush currents, such as motors, to prevent nuisance tripping. 5. Resettable Fuse Unlike traditional fuses that need to be replaced after blowing, resettable fuses can be reused multiple times. Also known as "polymeric positive temperature coefficient" (PPTC) fuses, they contain a material that will expand when there is an overcurrent, causing the resistance to increase and limit the current flow. When the current returns to normal, the material will shrink back to its original size, allowing the circuit to operate normally again. 6. Miniature/Automotive Fuse As the name suggests, miniature fuses are smaller in size compared to other fuses. They are typically used in electronic circuits and gadgets where space is limited.

Automotive fuses, on the other hand, are designed specifically for use in vehicles. They are usually color-coded for easy identification and have higher voltage ratings compared to other fuses. 7. HRC Fuse High Rupturing Capacity (HRC) fuses are used for high-voltage applications to protect electrical installations from heavy overloads, low overloads, and short circuits. They consist of a ceramic body with metal fusible elements and are designed to break quickly and without causing any damage in case of a fault. 8. SMD Fuse Surface-mount fuses, also known as "chip" fuses, are designed for use on printed circuit boards (PCBs). They are small in size and are soldered directly onto the PCB. SMD fuses are commonly used in consumer electronics, such as laptops and mobile phones, as they take up minimal space on the circuit board. 9. High Voltage Fuse As the name implies, high voltage fuses are designed to operate in circuits with higher voltage ratings, typically above 1000 volts. They come in various shapes and sizes, and some are designed to withstand high voltage surges, making them suitable for use in power distribution systems. 10. Electron Tube Fuse Electron tube fuses, also known as "metal-enclosed" fuses, are specifically designed for use in electron tubes, such as television sets and microwave ovens. They are constructed using a glass tube filled with sand and a small conductive strip. When there is an overcurrent, the strip will melt and cause the sand to fuse, breaking the circuit.

Fuse box

The fuse box is an important component in the electrical system of a building or vehicle. It serves as a safety mechanism to protect the circuits from overloads and short circuits, which could cause damage to the wiring or even lead to fires. The main function of a fuse box is to control the flow of electricity to different circuits in a building or vehicle. It consists of a series of fuses, which are small devices made of metal or glass that contain a thin strip of wire. When an excessive amount of current flows through the circuit, the wire inside the fuse heats up and melts, breaking the connection and cutting off the power supply to that specific circuit. In a building, the fuse box is usually located in the basement or utility room. It is connected to the main power supply coming from the grid and distributes electricity to the different circuits throughout the building. This includes the lights, appliances, outlets, and heating and cooling systems. Each circuit is connected to a separate fuse in the box, allowing for individual control and protection. The size and location of the fuse box may vary depending on the age and type of building. Older homes may have a fuse box with a limited number of circuits, while newer homes are equipped with a larger panel containing multiple fuses and circuit breakers. Commercial buildings and industrial facilities may have much larger fuse boxes to accommodate a higher demand for electricity. Similarly, in vehicles, the fuse box is responsible for distributing electricity to different circuits such as the lights, radio, and engine control systems. It is usually located under the hood, in the dashboard, or in the trunk of the vehicle. The size and complexity of the fuse box in a vehicle can vary depending on the make and model. One of the main advantages of a fuse box is its ability to protect the circuits from overloading. When a circuit is overloaded, it means that there is too much current flowing through it, which could cause the wires to overheat and potentially start a fire.

The fuse box prevents this by breaking the circuit and cutting off the power supply. In addition to preventing overloads, a fuse box also protects against short circuits. This occurs when there is a direct connection between two wires of different circuits, bypassing the fuse. In such cases, the fuse will blow and cut off the power to prevent damage to the wiring or appliances. While fuse boxes are effective in preventing electrical fires, they require regular maintenance to ensure they are functioning properly. It is essential to regularly check the fuses and replace them if they are blown or damaged. Over time, the fuses may also become loose or corroded, which can affect their performance and should be addressed promptly. In recent years, fuse boxes have been largely replaced by circuit breakers in new homes and vehicles. Circuit breakers work in a similar way but can be reset after tripping, eliminating the need for constant replacement of fuses. However, fuse boxes are still commonly found in older buildings and vehicles, and they continue to serve an important purpose in the protection of electrical systems.

Chapter3
Uses of fuse

1. Protecting Electrical Devices: One of the primary uses of fuses is to protect electrical devices from excessive current flow. When an electrical device is plugged into a socket, it draws a certain amount of current. If there is a fault in the device or the circuit, it can cause a surge of electricity that can damage the device or even cause a fire. A fuse acts as a barrier by breaking the connection and preventing the flow of electricity in such situations. It ensures that the device and the circuit are not subjected to more current than they can handle, thus protecting them from damage. 2. Overload Protection: In a residential

setting, the electrical system may have multiple appliances and devices connected to it. In such a scenario, there is a possibility that the total power drawn by all the devices may exceed the capacity of the electrical circuit, leading to an overload. This can cause damage to the wiring, and a fire hazard can arise. Fuses are designed to blow or melt when a circuit is overloaded, breaking the connection and stopping the flow of current. This protects the wiring and prevents any potential hazards. 3. Short Circuit Protection: Another common cause of electrical fires is a short circuit, which occurs when there is a direct connection between the hot and neutral wires in a circuit. This can happen due to damaged insulation or loose connections. Fuses are designed to detect such instances and interrupt the flow of current. They break the circuit, preventing the possibility of a fire and safeguarding the electrical system. 4. Protection from Power Surges: Power surges are sudden and temporary increases in voltage that can occur due to lightning strikes or utility grid fluctuations. These surges can damage sensitive electrical equipment such as computers, televisions, and appliances. Fuses act as a first line of protection by interrupting the flow of electricity in such situations. They can also prevent power surges from traveling to other circuits and damaging them. 5. Automotive Applications: Fuses are not only used in residential and commercial settings, but they also have a crucial role to play in automotive vehicles. Vehicles have various electrical systems, such as the engine, lights, and entertainment systems, which require different levels of current. Fuses are used to regulate the flow of electricity in each of these systems and protect them from damage in case of overloading or short circuiting. 6. Presence in Industrial Settings: Industrial settings, such as factories and warehouses, often have complex electrical systems and machinery. These systems are usually designed to handle a high amount of current, making them prone to overloading. Fuses are used in these settings to protect the machinery from damage due to an overload or a short circuit. They serve as an inexpensive but effective solution to prevent expensive damage or downtime. 7. Fire Safety: The primary purpose of fuses is to prevent fires caused by electrical faults. They play a crucial role in ensuring the safety of a building's occupants and the property. Fuses are installed at different points in an electrical circuit, making them capable of detecting and interrupting a fault anywhere in the system. They are the first line of defense against electrical fires and significantly reduce the risk of serious damage or injuries.

cartridge fuse

A cartridge fuse is a type of electrical safety device used to protect electrical circuits from overloading and potential damage. It is commonly found in the fuse box of homes and buildings, and also used in industrial and automotive applications. In this article, we will explore the history, workings, and benefits of cartridge fuses. History of Cartridge Fuses: The concept of a fuse dates back to 1864 when Thomas Edison patented the first electrical fuse. However, it wasn't until 1890 that the first cartridge fuse was invented by Nikola Tesla. This early version of the cartridge fuse used a glass tube filled with sand as the fuse element. It was later improved upon by Albert Schlemmer in 1894, who created a fuse with a porcelain body and tin-plated copper end caps. This design was the basis for modern cartridge fuses. How does a cartridge fuse work? A cartridge fuse consists of a metal or ceramic housing, a fuse element, and two end caps. The fuse element is made of a material with a low melting point, such as silver, copper, or aluminum. When an excessive amount of current flows through the circuit, it heats up the fuse element and causes it to melt, breaking the circuit and stopping the flow of electricity. This protects the circuit from overloading and causing damage to the electrical components. Benefits of Cartridge Fuses: 1. Higher current rating: One of the main advantages of cartridge fuses is their high current rating. They can handle higher amounts of current compared to other types of fuses, making them suitable for use in high-power applications. 2. Easy to replace: Cartridge fuses are designed to be easy to replace. They can be easily removed from the fuse holder and replaced with a new one without the need for any special tools. 3. Reliable protection: The design of cartridge fuses makes them more effective at protecting circuits from overloading. The fuse element is enclosed in a housing, making it less likely to be affected by dust, debris, or moisture. This ensures reliable protection for electrical circuits. 4. Space-saving: Cartridge fuses are a space-saving alternative to other types of fuses. They are compact in size and can be mounted closer to the electrical panel, thus taking up less room in the fuse box. 5. Versatility: Another benefit of cartridge fuses is their versatility. They come in a variety of sizes and current ratings, making them suitable for use in a wide range of applications, from small household circuits to large industrial systems. Types of Cartridge Fuses: There are two main types of cartridge fuses: 1. Class H: This type of fuse is designed for use in heavy-duty applications and can handle high currents up to 600 amps. 2. Class K5 and K9: These fuses are commonly used in automotive and industrial applications and have a higher voltage rating compared to Class H fuses. Tips for Choosing and Replacing Cartridge Fuses: 1. Always choose the correct size and type of cartridge fuse for your circuit. Using a fuse with a higher current rating than necessary can be dangerous and may cause damage to the circuit. 2. Before replacing a fuse, make sure the power to the circuit is turned off to avoid any accidents. 3. Use caution when handling cartridge fuses as they can get hot during operation or replacement. It is recommended to use gloves or a fuse puller tool when handling them.

The end

www.ingramcontent.com/pod-product-compliance
Lightning Source LLC
Chambersburg PA
CBHW050034230526
45470CB00003B/1269